MILITARY VEHICLES

AIRCRAFT CARRIERS

MARTY GITLIN

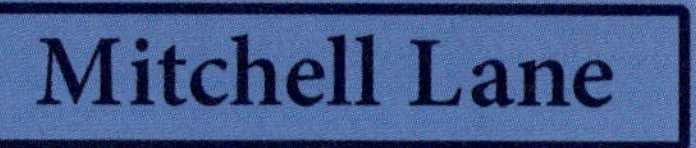

2001 SW 31st Avenue
Hallandale, FL 33009
www.mitchelllane.com

First Edition, 2021.

Author: Marty Gitlin
Designer: Ed Morgan
Editor: Morgan Brody

Little Mitchie is an imprint of Mitchell Lane Publishers.

Title: Military Vehicles: Aircraft Carriers / by Marty Gitlin
Description: Hallandale, FL :
Mitchell Lane Publishers, [2021]

Series: Military Vehicles
Library bound ISBN: 978-1-58415-238-5
eBook ISBN: 978-1-58415-239-2

Photo credits: Freepik.com, Freevector.com, cover: defense.gov, p. 5 U.S. Navy, p. 6 U.S. Navy, p. 7 public domain, p. 9 Canadian public domain, p. 10 U.S. Navy, p. 11 U.S. Navy, p. 13 U.S. Navy, p. 14 U.S. Navy, p 15 U.S. Navy, p. 17 British Navy, p. 18 U.S. Navy, p. 19 U.S. Navy

CONTENTS

Words in **bold** can be found in the Glossary.

1
BATTLING BACK

The date was December 7, 1941. The U.S. aircraft carrier *Enterprise* was headed to Pearl Harbor. That was a Navy base in Hawaii. The country was about to enter World War II.

Rescuers work to find survivors from the *USS Virginia*.

A horrible sight awaited the crew. They saw American ships destroyed. More than 2,000 people were dead. They had been killed in an **attack** by Japanese planes.

Those on the *Enterprise* knew two things. They too might have been dead had they been there that morning. And they had to help America win the war.

The crew went to work. Four U.S. carriers were sunk by Japanese planes and **submarines**. But not the *Enterprise*. It sank a Japanese sub a day after Pearl Harbor. It carried planes into battle. Its **bombers** wiped out enemy ships.

The carrier helped win one battle after another. It remains one of the greatest war ships ever. The *Enterprise* earned twenty battle stars. That is more than any U.S. ship during the war.

A U.S. Navy Torpedo Squadron prepares to launch from the *USS Enterprise* in June 1942.

FAST FACT

THE 'OTHER' *ENTERPRISE*

The *Enterprise* is a famous spaceship. But this one was not real. It was manned by the crew in the classic 1960s TV show *Star Trek*. Some believe it was named after the aircraft carrier *Enterprise*. But that has never been proven.

A scene from the *Star Trek* episode "Spock's Brain" that aired September 20, 1968.

2

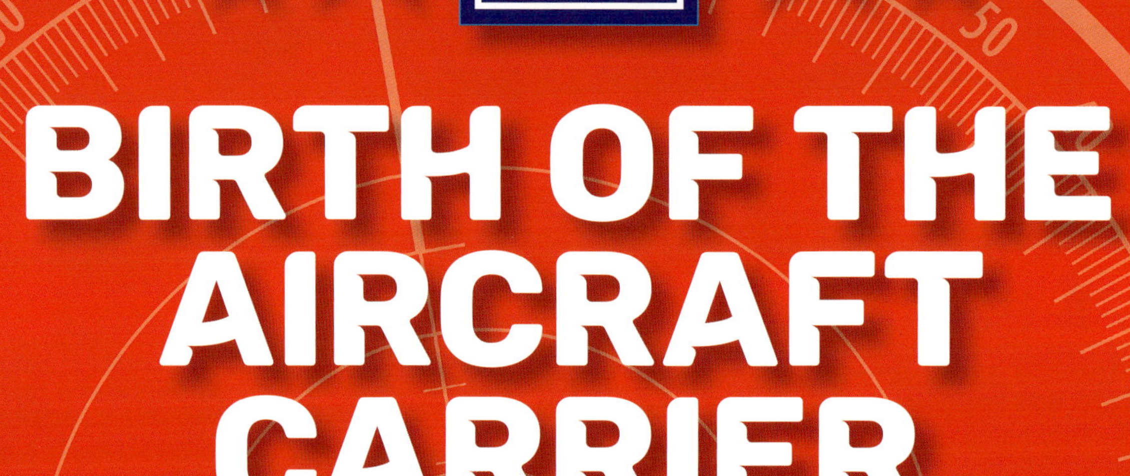

BIRTH OF THE AIRCRAFT CARRIER

Imagine the aircraft carrier when the aircraft was a balloon! Austria used the first craft holding a hot air balloon. That country tried to launch bombs at Italy from the balloon. But it failed.

The American army used floating **barges** during the Civil War. The boats were tied to balloons with cameras. They spied on enemy soldiers from far away.

The French Navy built the first aircraft carrier in 1911. But planes could not launch from the ship. They were lowered into the water before flying. The British began sending planes from flat carriers during World War I.

The S.S.Z. 59 lands on the deck of Britain's largest aircraft carrier the *H.M.S. Furious* during World War I.

Countries turned existing ships into carriers in the 1920s. They were not used in war again until the late 1930s. That is when their planes bombed targets in World War II.

Commanders talk on the *USS Wasp* in August 1942 with Douglas SBD-3 Dauntless scout bombers on the flight deck.

FAST FACT

PLANTING THE SEEDS OF SUCCESS

The first U.S. Navy aircraft carrier was the *USS Langley*. But it was not built from scratch. It had been the *USS Jupiter* and converted into a carrier in 1922. It was 532 feet long and could travel nearly 18 miles an hour. The *Langley* could carry a crew of 468 people.

3 HOW THEY SERVE

Many films of carriers in action have been shown on TV. Those moments can be exciting. They show warplanes zipping off from carriers to their targets.

The U.S. Navy used aircraft carriers to win many World War II battles. The ships often carried about one hundred planes.

The aircraft built up speed from the **deck**. They would then take off. The planes flew through the sky. They dropped bombs on enemy ships. They often flew far enough to bomb cities or towns. They also hit factories in which the enemy was building weapons.

In 1943, a bomber waves-off during flight operations aboard the aircraft carrier *USS Bunker Hill* in the Caribbean.

Soon bigger Navy carriers were built. Those 1950s ships had huge **hangars** and flight decks. They were later powered by **nuclear** energy. They launched planes faster than ever.

Aircraft carriers have indeed changed. They have been used to fill the needs of the U.S. military.

A Grumman F9F-2 Panther jet goes up the elevator on the *USS Bon Homme Richard* to the flight deck where it was gassed and armed for a Thanksgiving Day strike in Korea in November 1952.

FAST FACT

MOVING RIGHT ALONG

How fast are modern U.S. aircraft carriers? The new ships travel 35 miles per hour. That is twice as fast as the first Navy carriers of the 1920s. They are more than 1,000 feet long and support up to 5,000 people! They are called Ford Class carriers. The ships were named after 1970s U.S. President Gerald Ford.

4 WHAT MAKES THEM GREAT?

No other ship can do what aircraft carriers can do. They counter or start air attacks. They battle enemy ships or submarines. They send planes to strike or spy on land.

Carriers can also be used to help people. They can deliver food and supplies to those hit by floods or earthquakes.

USS John C. Stennis is pictured from the Royal Navy frigate *HMS St. Albans* as a helicopter transfers goods between the two ships.

The **vessels** can carry any kind of plane. That is truer today than ever. Carrier size and strength are very important. They help carriers handle larger planes with new **technology**.

There is one big problem. Aircraft carriers are huge targets. They are also slow. U.S. Navy crafts can hold 5,000 sailors. They are barely armed. That many lives could be lost if one is sunk.

Carriers are in danger from enemy planes, ships, and **missiles**. But the Navy believes the **benefits** outweigh the risks.

Several airplanes fly in formation over the *USS Ronald Reagan* in 2019.

The *USS Nimitz* in the Gulf of Oman.

FAST FACT

NAMED AFTER NIMITZ

The U.S. Navy features ten Nimitz aircraft carriers. They are nuclear-powered vessels. The class of ships is named after Chester W. Nimitz. He was a fleet admiral who helped defeat Japan in World War II.

SAILING INTO THE FUTURE

U.S. aircraft carriers have helped win wars in the past. But it is hoped that they will help keep peace in the future.

The military plans to use them in many roles. They can rush aid to disaster victims. Millions of Americans can be in danger. Volcanoes can strike in Hawaii. Earthquakes often hit California. Hurricanes **batter** the eastern and southern states.

Future fleets will not just carry planes. They will also house **helicopters** and **drones**. Those craft can aid people in many ways. They can help the Navy deliver supplies. They can bring food to people around the world.

Carriers can prevent war. They show military strength. They can stop nations from risking attacks. The hope is that these ships will be used to save rather than hurt people.

FAST FACT

ONE COSTLY CARRIER!

A modern aircraft carrier is sure not cheap to build. One example is the *USS Gerald R. Ford*. That was the lead vessel of the Navy fleet in 2018. It is the most expensive warship in U.S. history. It cost more than 13 billion dollars to build!

What You Should Know

- Aircraft carriers transport planes that can do many missions.
- The technology is more than one hundred years old.
- Aircraft carriers have been used by the U.S. Navy in times of war and peace.
- The peak of carrier use was during World War II.
- Carriers can hold as many as 5,000 sailors.
- Slow, heavy carriers would be easy to attack.
- Spy balloons were attached to the earliest U.S. carriers.

Glossary

attack
Military strike against an enemy

barge
Boat with a flat bottom usually for hauling goods in harbors

batter
To hit (something or someone) forcefully in a way that causes damage or injury

benefit
Something that does good to a person or thing

bomber
Plane that drops bombs on targets

deck
Floor of a ship

drone
Aircraft without a pilot controlled by radio signals

hangar
Shelter for housing and repairing aircraft

helicopter
Aircraft supported in the air by propellers

missile
An object launched at a target from a long distance

nuclear
Related to highly destructive weapons

submarine
Naval vessel that operates underwater

technology
Use of science to solve problems

vessel
A watercraft bigger than a rowboat

Find Out More

Ackerman, Melissa. *Boats and Ships for Kids*. CreateSpace Independent Publishing Platform, 2016.

Doeden, Matt. *Aircraft Carriers*. Minneapolis, MN: Lerner Publishing, 2005.

Roza, Greg. *The Incredible Story of Aircraft Carriers*. New York, N.Y.: Rosen Publishing, 2004.

Websites

Ducksters: This website offers interesting information for kids about aircraft carriers.
https://www.ducksters.com/history/world_war_ii/aircraft_carriers_in_ww2.php

Britannica Kids (Navy): Learn all about the U.S. Navy here.
https://kids.britannica.com/kids/article/navy/353522

American History for Kids: This site provides fun facts about World War II sea battles.
https://www.americanhistoryforkids.com/world-war-ii-victory-pacific/

America's Navy: View hundreds of photos of aircraft carriers through the years.
https://www.navy.mil/viewGallery.asp?id=10

Index

About the Author

Martin Gitlin is a freelance author based in Cleveland, Ohio. He has written about 150 books since 2006. Most have been targeted for young students. He has maintained a strong interest in military vehicles such as aircraft carriers. Martin worked as a newspaper journalist from 1991 to 2002. He won more than 45 awards during that time. Included was a first place for general excellence from Associated Press in 1996. That organization also selected him as one of the top four feature writers in Ohio in 2002.